A CONCISE INTRODUCTION TO BAYES' THEOREM

KELLY J. KIRKLAND

IN MEMORY OF PIERRE-SIMON LAPLACE

(1749—1827)

Introduction

You just woke up. You haven't heard the weather forecast; you haven't even looked outside yet. I ask you, "What is the probability it will rain today?" You answer, "I don't know. It doesn't rain very often this time of year. Maybe 15 percent."

But then you look out the window. Ominous black rain clouds are moving in. "Wait a minute," you say. "More like 95 percent."

Or you look out the window and the sky is completely cloudless. "Wait a minute," you say. "More like 5 per cent."

This is important, so let's think carefully about what just happened. In both cases you took your initial personal belief as to the probability of something and then modified that subjective belief when you got a new piece of information. But how did you know how large a modification to make? Did you simply replace one guess with another?

What if there were a way of making these modifications using a mathematical formula instead of just guessing? There is. And what if you could both prove the validity of the formula and use the formula to make calculations for everyday scenarios without needing calculus or indeed any mathematics beyond simple algebra? You can.

Chapter One: What is Bayes' Theorem? [1]

Let's take a more complex example than simply whether you think it is going to rain today. Consider this situation: Your company uses a machine to make widgets. At the end of each shift, a maintenance crew performs routine maintenance on the machine to keep it in top working condition. To do this, the maintenance crew must open a valve on the machine. Once the routine maintenance has been completed, the maintenance crew is supposed to close the valve. Restarting the machine with the valve open can cause the machine to explode. Nevertheless, operating crews have reported to management that 5% of the time they find the valve open when they arrive to start their shift.

There are other reasons why the widget machine might explode, but that one is on everyone's mind today. Two days ago, one operating crew finished its shift and left. The maintenance

---

[1] Current punctuation practices dictate that the name of the theorem be written "Bayes's Theorem." However, most people still write "Bayes' Theorem," and we will continue that practice in this book. Feel free to add another "s" if it bothers you. In other words, it can be "Bayes' Theorem" or "Bayes's Theorem," but not "Baye's Theorem."

crew arrived and performed its work while the next operating crew stood next to the machine, waiting to start its shift. The maintenance crew finished its work and hung around, joking with the operating crew as they restarted the machine. Unfortunately, there was a huge explosion, which obliterated the machine and killed every member of the maintenance and operating crews.

Today your boss called you into her office and gave you the task of conducting an investigation into the tragedy. Look at the records, talk to people, do whatever research you want—then write up a report to management. Tell them what you think happened.

There is one surviving worker, who was standing some distance away from the machine at the time. He tells you he watched the operating crew start up the machine. He says that he saw that the valve was still open but did not have time to warn the crews prior to the explosion. You decide to run an experiment. It turns out that from his vantage point on the day of the accident, this witness can determine whether the valve on a widget machine is open or shut with a high degree of accuracy; specifically, he will correctly say that the valve is open 90% of the times it actually is open and will incorrectly say it is open only 10% of the times it actually is shut.

The widget machine itself was so badly damaged that it was impossible to learn anything about the cause of the explosion from examining what was left of it. You conduct the best investigation you can, but in the end this is all you've got:

(i)   Company records show that 5% of the time operating crews report that the maintenance crews leave the valve open when they have finished their work.

(ii)   The sole surviving eyewitness says he saw the valve open right before the explosion. You believe he is right 90% of the time.

When you started your investigation, you thought there was only a 5% chance that the valve was open based on company records. But talking to the eyewitness and testing his ability to tell whether the valve is open from his position at the time of the explosion caused you to raise your belief as to the probability that the valve was open from 5% to 90%. Pretty straightforward, no?

No. Bayes' Theorem teaches us how to modify our existing subjective beliefs as to the probability of something when we get a new piece of evidence. In this situation, Bayes' Theorem tells us that the *rational* response to the eyewitness evidence would be to raise your belief as to the probability that the valve was open from 5% to approximately 32%—less than one chance in three. How in the world could that be?

Chapter Two—Deriving Bayes' Theorem

Let's define some terms. In this book we will explore our subjective beliefs about how likely it is that certain propositions are true. Let's call those propositions "hypotheses" and label them "H." In Chapter One our hypothesis was that the valve was open immediately before the explosion.

We want to evaluate how our subjective belief about the probability that a hypothesis is true should change when we obtain a new piece of evidence relevant to the hypothesis.[2] Let's label the new piece of evidence "E." In Chapter One the new evidence was the testimony of the surviving eyewitness that the valve was open immediately before the explosion.

We therefore have a hypothesis (H) and a new piece of evidence (E). If both the hypothesis is true and the new piece of evidence exists, we will write that as: (H & E). Furthermore, we

---

[2] A new piece of evidence is *relevant* to the hypothesis if it would cause a rational person to change his/her belief as to the probability that the hypothesis is true.

will let a lower case "p" stand for "probability."[3]   Thus, the probability that the hypothesis is true is: p(H).   Similarly, the probability that the new piece of evidence exists is: p(E).   The probability that both the hypothesis is true and the new piece of evidence exists is: p(H & E).

Suppose we now ask, "What is the probability of H <u>given</u> <u>E</u>?"   That is, if we assume that E exists, what is the probability that H is true?   "H given E" is: (H|E); "the probability of "H given E" is: p(H|E).

To answer this question, we must look only at the set of all situations where E exists and then ask what percentage of those situations also include H being true. Thus:

$$(H \mid E) = \frac{(H \& E)}{E}$$

and

$$p(H \mid E) = \frac{p(H \& E)}{p(E)}$$

[3]   You should be aware that some authors use an upper case "P" or a combination "Pr" to stand for "probability."

One more insight and then watch the dominos fall. (H & E) is the same thing as (E & H). In other words, saying "H is true and E exists" is the same thing as saying "E exists and H is true."

We already know that:

$$p(H \mid E) = \frac{p(H \& E)}{p(E)}$$

Multiplying both sides by p(E), we get:

$$p(H \mid E)\, p(E) = p(H \& E)$$

Similarly: What is p(E|H)? Here we must look only at the set of all situations where H is true and then ask what percentage of those situations also include E as existing. Proceeding as before:

$$(E \mid H) = \frac{(E \& H)}{H}$$

and

$$p(E \mid H) = \frac{p(E \& H)}{p(H)}$$

Multiplying both sides by p(H), we get:

$$p(E \mid H)\, p(H) = p(E \& H)$$

But if (H & E) is the same thing as (E & H), it follows that their probabilities are the same:

$$p(H \& E) = p(E \& H)$$

It then follows that:

$$p(H \mid E)\, p(E) = p(E \mid H)\, p(H)$$

Dividing both sides by p(E) we get:

$$p(H \mid E) = \frac{p(H)\, p(E|H)}{p(E)}$$

And that's Bayes' Theorem![4]

Stay with me just a little bit longer. We can tweak this formula to make it more useful when we make calculations. The denominator on the right side of the equation is p(E). There are two possibilities if E exists—(i) E exists and H is true, and (ii) E exists and H is not true. "H is not true" is: ~H. Thus, the two possibilities are (i) (E & H), and (ii) (E & ~H).

It follows that:

$$p(E) = p(E \& H) + p(E \& \sim H)$$

We've already shown that:

---

[4] Note that p(E) cannot equal 0.

$$p(E \& H) = p(H) \, p(E|H)$$

And by similar reasoning:

$$p(E \& {\sim}H) = p({\sim}H) \, p(E|{\sim}H)$$

To get the probability of E, we simply add the two probabilities:

$$p(E) = [p(H) \, p(E \mid H)] + [p({\sim}H) \, p(E \mid {\sim}H)]$$

We're almost there. A few minutes ago we expressed Bayes' Theorem as:

$$p(H \mid E) = \frac{p(H) \, p(E|H)}{p(E)}$$

Now we know that:

$$p(E) = [p(H) \, p(E \mid H)] + [p({\sim}H) \, p(E \mid {\sim}H)]$$

Bayes' Theorem thus becomes:

$$p(H \mid E) = \frac{p(H) \, p(E|H)}{[p(H) \, p(E \mid H)] + [p({\sim}H) \, p(E|{\sim}H)]}$$

where:

p = probability

H = hypothesis

E = (new) evidence

| = given

$\sim$ = not

That's the formula for Bayes' Theorem we are going to use in this book. You will get to know it very well.

Chapter Three—Applying Bayes' Theorem (Example 1)

Let's go back to Chapter One. H represents the hypothesis that the valve was open and E represents the surviving eyewitness' testimony that he believes he saw that the valve was open. Before the eyewitness testimony, company records indicated that the probability that the valve was open is 0.05, or 5%. Then you talked to the eyewitness, who said that the valve was open immediately before the explosion. We also know that there is a 90% probability (0.90) that the eyewitness' testimony is correct if the valve was indeed open and a 10% probability (0.10) that his testimony is incorrect if the valve was closed. That testimony increases the probability that the valve was open from 5% to X%. What is X?

Applying Bayes' Theorem:

$$p(H \mid E) = \frac{p(H)\,p(E|H)}{[p(H)\,p(E \mid H)] + [p(\sim H)\,p(E|\sim H)]}$$

$$p(H) = 0.05$$

$$p(\sim H) = 0.95$$

$$p(E|H) = 0.90$$

$$p(E|\sim H) = 0.10$$

$$p(H \mid E) = \frac{(0.05)(0.90)}{[(0.05)(0.90)] + [(0.95)(0.10)]}$$

$$p(H \mid E) = \frac{0.045}{0.045 + 0.095}$$

$$p(H \mid E) = \frac{0.045}{0.140}$$

$$p(H|E) = 32\%$$

Please take all the time you need to make sure you understand this calculation.

The values for p(H) and p(~H) come from the pre-explosion company records. These are the probabilities *before* considering the new piece of evidence. Notice that if the records show that the maintenance crew improperly leaves the valve open 5% of the time, it follows that they properly close it 95% of the time. It is either open or closed when they turn the machine over to the next operating crew—there are no other possibilities. So:

$$p(H) = 0.05$$

$$p(\sim H) = 0.95$$

The values for p(E|H) and p(E|~H) come from the test of the accuracy of the eyewitness's testimony that you ran after the explosion. If the valve is indeed open—that is, given H—the eyewitness will say that it was open (that is, E) 90% of the time. If the valve was not open—that is, given ~H—the eyewitness will say (incorrectly) that the valve was open (that is, E) 10% of the time. So:

$$p(E|H) = 0.90$$

$$p(E|\sim H) = 0.10$$

The rest is just plugging these values into the formula for Bayes' Theorem that we derived in Chapter Two.

The testimony of the eyewitness should increase the subjective probability that the valve was open at the time of the explosion from 5% to 32%—not to 90% or even something over 50%. How can this be? Our minds intuitively overweight the 90% probability that the eyewitness can correctly see that the valve is open *during the 5% of the times that it is open* and ignore (or at least substantially underweight) the 10% probability that the eyewitness incorrectly "sees" that the valve is open *during the 95% of the times that it is not open.* If the eyewitness looks at the valve 1,000 times, we started out believing that on average it will be open 50 times and closed 950 times. The eyewitness will correctly say that valve is open 45 of the 50 times that is indeed open. But he will also incorrectly say that it is open 95 of the 950 times that it is in fact closed.[5] Our intuition usually ignores these "false positives." This is a common fallacy in our everyday thinking, one that we can correct by using Bayes' Theorem.

---

[5] Thus, he is correct 45 times out of the 140 times [45 + 95] he says the valve is open, or approximately 32% of the time.

Chapter Four—Applying Bayes' Theorem (Example 2)

Assume you are a doctor in a small town. Everyone in town is on edge today. Last night the lead story on the local TV news was about a new virus that has made a number of people ill in the state capital. Not just ill, but terminally ill, as in there is no known cure.

Experts aren't sure yet, but transmission of the virus may be airborne. In any event, the working assumption is that everyone in the capital has already been exposed to the virus. The only good news is that although it may be very easy to become exposed, it is apparently difficult to become infected.[6] In fact, indications are that only 3 people out of every 1,000 people exposed to the virus will become infected. But still—no known cure, inevitably fatal— this is scary stuff.

---

[6] For our purposes, a person is *exposed* to the virus when the virus comes into contact with his or her body. If this virus is indeed transmitted through the air, a person would probably become exposed when s/he inhales the virus. A person is *infected* by the virus when his/her immune system does not destroy the virus and instead the virus invades the person's bodily tissues and causes an illness— in this case, a fatal illness.

Your first few patients come in this morning with routine illnesses, nothing out of the ordinary in your practice. But just before lunch, one of your oldest friends comes in without an appointment. You've known him since you were kids; you spent a nice evening at his home with his wife and two small kids just last week, before he left for a convention in the state capital. You squeeze him in, of course.

He tells you he's not feeling well at all. You examine him, but you're not sure what's wrong with him. You are well aware that he was in the state capital only a few days ago; you assume he was exposed to the new virus. He is frightened by the news reports from the capital. You are too, but you don't tell him that.

You reach for the medical test for the new viral illness, the one you got a couple of days ago. You remind him that for every 1,000 people who are exposed to the virus, only 3 become infected. But for peace of mind—yours as well as his—you run the test.

As you learned in your medical training, ideally you want a medical test for an illness to be positive if and only if the patient has the illness and to be negative if and only if the patient doesn't have the illness. That is, you'd prefer not to have any "false positives" or "false negatives." In the real world, however, you usually have a trade-off. The harder they work to design a test that comes back positive when the patient has the illness—that is, the test is highly "sensitive"—the more likely they will overshoot the mark and get positive results for some patients who don't actually have the illness. Conversely, the harder they work to design a test that comes back negative when the patient doesn't have the illness—that is, the test is highly "specific"—the more likely they will overshoot the mark and get negative results for some patients who do actually have the illness.

The test you are holding in your hand is very good. If the patient being tested is infected, there is a 99% probability that the test will come back positive. If the patient isn't infected, there's only a 2% chance the test will come back positive. You run the test and send it to the lab; it will take a couple of days to get the result.

Two days later, your friend is much sicker; he is now hospitalized. You get the test results back—Positive. You immediately go to the hospital to break the news to your friend. What are you going to tell him?

You shouldn't tell him anything before applying Bayes' Theorem. Let H be the hypothesis that your friend has the viral illness. Let E be the positive test result. What is the probability that your friend has the viral illness given that he tested positive?

$$p(H \mid E) = \frac{p(H)\,p(E|H)}{[p(H)\,p(E \mid H)] + [p(\sim H)\,p(E|\sim H)]}$$

$$p(H) = 0.003$$

$$p(\sim H) = 0.997$$

$$p(E|H) = 0.990$$

$$p(E|\sim H) = 0.020$$

Notice in passing that p(E|H) + p(E|~H) does not always equal 1.000 [although p(H) + p(~H) does].

Plugging in these values:

$$p(H \mid E) = \frac{(0.003)(0.990)}{[(0.003)(0.990)] + [(0.997)(0.020)]}$$

$$p(H \mid E) = \frac{0.00297}{0.00297 + 0.01994}$$

$$p(H \mid E) = \frac{0.00297}{0.02291}$$

$$p(H|E) = 13\%$$

You should tell your friend that you believe that his chances of being infected did increase substantially as a result of testing positive, from 3 in a 1,000 to 130 in a 1,000.

But that's still only 13%. Sure, you're going to take his case seriously—very seriously. Supportive treatment; test for other illnesses; watch him like a hawk. And tell him that there are real grounds for hope, because despite the 99% and 2% statistics surrounding the test, the relative rarity of the illness means that there are going to be a very substantial number of "false positives." In fact, Bayes' Theorem tells us that there will be many more false positives than true positives.

Chapter Five—Applying Bayes' Theorem (Example 3)

You just had a talk with your neighbor. Your nosy neighbor, the one who just loves to gossip. "I saw your husband yesterday. I was passing by a restaurant over in the East End. He was having lunch with a very attractive woman. I was going to say hello but they were deep in conversation. But I'm sure he told you all about it."

You've been married for 10 years; two kids. Sure, things between the two of you aren't what they used to be before the kids came along, but that's normal, right? There's probably a perfectly reasonable explanation that your husband, who usually grabs a sandwich and eats at his desk, went all the way across town to have lunch with an attractive woman. He's been so busy lately he probably just forgot to tell you about it. There's no way he could be having an affair. Absolutely no way. Right?

Bayes' Theorem can help here:

$$p(H \mid E) = \frac{p(H)\,p(E|H)}{[p(H)\,p(E \mid H)] + [p(\sim H)\,p(E|\sim H)]}$$

You need to come up with values for four probabilities, values that you think are at least reasonably reliable. The four probabilities you need are:

$$p(H)$$

$$p(\sim H)$$

$$p(E|H)$$

$$p(E|\sim H)$$

Let's take them one at a time.

Here H is the proposition that your husband is having an affair. Let's start by defining E as the fact that your husband had lunch yesterday with an attractive woman at a restaurant in the East End.[7] We will refine our definition of E in a moment, but for now you are going to calculate $p(H \mid E)$; that is, the probability that your husband is having an affair given that he had lunch yesterday with an attractive woman at a restaurant in the East End.

That makes p(H) the probability you would assign to the proposition that your husband is having an affair without taking

---

[7] Let's assume that you believe your neighbor—she may be a gossip but she's not a liar.

into account this latest piece of news. You go online. There are all kinds of statistics about the prevalence of marital infidelity. And here you have to be careful. What exactly do you want to know? Not the probability that a married man will cheat *at some point during his marriage.* Instead you want to know what percentage of married men are having an affair at any given time because you want to know if your husband is having an affair right now, not whether he has had an affair in the past or will have an affair at some future point during your marriage.[8]

After doing some research you find a range of estimates for the probability that a married man is having an affair at any given time. If you can find it, even better would be an estimate for men who have been married for 10 years. Or for men who have been married for 10 years and have children. The closer to your situation the better.

You still find a range of estimates. Now you face the first hurdle—can you find a probability estimate that seems reasonable

---

[8] Actually, you would love to know all those probabilities, but first things first!

enough to you to use in your calculation? Because if you can't plug in an estimate for p(H) that you have some confidence in— maybe not absolute certainty but an estimate that sounds "about right" to you—you can't use Bayes' Theorem.[9]

Let's say you settle on a value for p(H) of 0.05; that is, at any given time 5% of married men similar to your husband are having an affair.[10] That makes p(~H) 0.95. What you want to know now is how the fact that your husband had lunch yesterday with an attractive woman at a restaurant in the East End affects the probability that he is having an affair—p(H|E).

The next probability you need is p(E|H). Be careful here. It is easy to confuse p(E|H) and p(H|E). Here p(E|H) is the probability that your husband would have lunch with his paramour *given that (assuming that) he is having an affair.* Not necessarily at that particular restaurant—there are lots of restaurants in the

---

[9] Actually, there is an important caveat to that general statement, but we'll talk later about the phenomenon called "washing out the priors."

[10] By the way, I have no idea if that is a good number, so *please* don't jump to any conclusions about the fidelity of your husband or about men in general based on this example.

East End and it really doesn't matter which one he picked. Nor does it matter that they were in the East End, as opposed to somewhere else significantly distant from his office. Nor does it matter that it was yesterday, as opposed to some other day.

So more precisely—what is the probability that your husband would have lunch with his paramour at a restaurant significantly distant from his office *if* he is indeed having an affair? You think it's pretty high; the guy is with you every evening, so if he's cheating on you it has got to be during the daytime and meeting her for lunch makes perfect sense. You set $p(E|H)$ equal to 0.85, or 85%.

The last probability you need is $p(E|{\sim}H)$; that is, the probability that your husband would have lunch with an attractive woman at a restaurant significantly distant from his office if he is *not* having an affair. That probability is not necessarily 0.15, that is, $1.00 - 0.85$. As we have already seen, the sum of $p(E|H)$ and $p(E|{\sim}H)$ is not necessarily 1.00 (although the sum of $p(H)$ and p(~H) must be 1.00—your husband is either having an affair or he is not having an affair; there is no third possibility).

In fairness to your husband, p(E|~H) is not 0. There are some scenarios that would account for yesterday's lunch that do not involve an affair. Maybe she is an important client or an outside salesperson. But you're not in a generous mood, so you set p(E|~H) at 0.10, or 10%.

Hmm ... p(E|H) is 85% and p(E|~H) is only 10%. Let's run the numbers so that you have them when the cheating bastard gets home tonight:

$$p(H \mid E) = \frac{p(H)\,p(E|H)}{[p(H)\,p(E \mid H)] + [p(\sim H)\,p(E|\sim H)]}$$

$$p(H) = 0.05$$

$$p(\sim H) = 0.95$$

$$p(E|H) = 0.85$$

$$p(E|\sim H) = 0.10$$

Plugging in these values:

$$p(H \mid E) = \frac{(0.05)(0.85)}{[(0.05)(0.85)] + [(0.95)(0.10)]}$$

$$p(H \mid E) = \frac{0.0425}{0.0425 + 0.0950}$$

$$p(H \mid E) = \frac{0.0425}{0.1375}$$

$$p(H|E) = 31\%$$

The fact that your husband had lunch yesterday with an attractive woman at a restaurant significantly distant from his office should increase your subjective belief as to the probability that your husband is having an affair from 5% to 31%. That's a significant increase, but you should still believe that it is more likely that he is *not* having an affair than that he is.

Chapter Six—Applying Bayes' Theorem (Example 4)

This is frustrating. You are the chief operating officer of a company that makes electric thing-a-ma-jigs. Thing-a-ma-jigs are notoriously temperamental. It is not always as simple as taking one out of the box and plugging it in. Sometimes you have to do some on-the-spot repairs to get it to work.

As usual, you get what you pay for. Although they look identical from the outside, there are two kinds of thing-a-ma-jigs, one more expensive than the other. You pride yourself on the fact that every thing-a-ma-jig is tested before it leaves your factory, and the ones that need tinkering get it. The more expensive ones will last longer, to be sure, but they will all work when the customer takes one out of the box.

Or so you thought. This morning a customer returned one because it doesn't work. You look up the sales records—this customer bought an equal number of expensive and inexpensive thing-a-ma-jigs and has no idea which box this one came out of.

Normally the serial number would tell you, but in a fit of anger the customer tried to fix the thing-a-ma-jig himself and succeeded only in obliterating the face of the product. All you have to work with is your knowledge that (i) 50% of your sales to this customer are expensive thing-a-ma-jigs and 50% are inexpensive ones, and (ii) the more expensive kind have a "failure" rate of 12%— that is, out of every 25 you can expect that 3 will require tinkering to get them to work—while the less expensive kind have a "failure" rate of 28%, that is, out of every 25 you can expect that 7 will require tinkering.

You could disassemble this particular thing-a-ma-jig to see which kind it is. You could, but you really don't want to. Can Bayes' Theorem help here? Sure it can.

This formula should look familiar by now:

$$p(H \mid E) = \frac{p(H)\,p(E|H)}{[p(H)\,p(E \mid H)] + [p(\sim H)\,p(E|\sim H)]}$$

Let's define some terms. Let H be the hypothesis that the thing-a-ma-jig is one of the expensive ones. Let E be the fact that the thing-a-ma-jig didn't work when the customer took it out of the box and plugged it in. You want to know the probability that the thing-a-ma-jig is one of the expensive thing-a-ma-jigs given that it didn't work when the customer took it out of the box and plugged it in, which is p(H|E).

To apply Bayes' Theorem, we need to know p(H), which is the probability that this is an expensive thing-a-ma-jig before you learned it doesn't work. We know that—50%, or 0.50. The probability that it is an inexpensive thing-a-ma-jig [p(~H)] is therefore also 50%, or 0.50.

We also need to know p(E|H), which is the probability that a thing-a-ma-jig would not work without tinkering (which this one apparently somehow didn't get) if it is an expensive one. We know that—12%, or 0.12. What about p(E|~H)? That is the probability that a thing-a-ma-jig would not work if it is an inexpensive one. We know that also—28%, or 0.28.

You may be able to do this one in your head, but let's run the numbers:

$$p(H \mid E) = \frac{p(H)\, p(E|H)}{[p(H)\, p(E \mid H)] + [p(\sim H)\, p(E|\sim H)]}$$

$$p(H) = 0.50$$

$$p(\sim H) = 0.50$$

$$p(E|H) = 0.12$$

$$p(E|\sim H) = 0.28$$

$$p(H \mid E) = \frac{(0.50)(0.12)}{[(0.50)(0.12)] + [(0.50)(0.28)]}$$

$$p(H \mid E) = \frac{0.06}{0.06 + 0.14}$$

$$p(H \mid E) = \frac{0.06}{0.20}$$

$$p(H|E) = 30\%$$

You should believe that there is a 30% probability that the thing-a-ma-jig that the customer is returning is an expensive one.

Chapter Seven—Applying Bayes' Theorem (Example 5)

Let's do one more example, but this time you do the math.

You are the President of the United States. There have been many mass shootings recently, and they seem to be occurring with increasing frequency. There is a lot of pressure on you to do something.

A maverick psychologist has recently published some intriguing research. He has come up with a new algorithm that purports to predict which people have a serious likelihood of becoming a mass murderer, people who obviously need to watched very closely. His research indicates that if a person is indeed one of these potential mass murderers, this algorithm will flag him with an astounding 99.9% accuracy. Equally important, if that person is not a potential mass murderer, the algorithm will flag him incorrectly only 1% of the time. He estimates that there are around 3,250 people who have a serious likelihood of becoming a mass murderer out of a total population of 325 million.

This is clearly interesting research, but as always there are drawbacks. To apply this algorithm requires a massive collection of personal information from the entire population. The program is going to be highly controversial and will probably face serious and expensive legal challenges. If you decide to go down this road, you want to make sure that it is worth the trouble.

Before applying the algorithm, an estimated 1 in 100,000 people (0.001% or 0.00001 of the total) are potential mass murderers.[11] Even assuming it works as advertised, how much will the algorithm help? How likely is it that someone flagged by the algorithm has a serious likelihood of becoming a mass murderer?

The next page is blank. You can use it to do your calculations. The answer is on page 36—but try to work out the problem yourself first.

---

[11] Again, I just made up this statistic for purposes of illustration. Do not rely on it for any purpose other than learning how to use Bayes' Theorem!

$$p(H \mid E) = \frac{p(H)\, p(E|H)}{[p(H)\, p(E \mid H)] + [p(\sim H)\, p(E|\sim H)]}$$

$$p(H) = 0.00001$$

$$p(\sim H) = 0.99999$$

$$p(E|H) = 0.999$$

$$p(E|\sim H) = 0.010$$

$$p(H \mid E) = \frac{(0.00001)(0.999)}{[(0.00001)(0.999)] + [(0.99999)(0.010)]}$$

$$p(H \mid E) = \frac{0.00000999}{0.00000999 + 0.0099990}$$

$$p(H \mid E) = \frac{0.00000999}{0.01000989}$$

$$p(H \mid E) = 0.00099800 = 0.0998\%$$

If the algorithm flags someone as a potential mass murderer, it is rational to believe there is still less than 1 chance in 1,000 that the person demonstrates a serious likelihood of becoming a mass murderer—which is down substantially from 1 chance in 100,000. On the other hand, the algorithm will flag more than 3,250,000 people to pick out the 3,250 people of interest.

Chapter Eight—Fun Facts About Bayes' Theorem

So far we have learned the basic idea behind Bayes' Theorem, derived its formula, and learned how to apply it in everyday situations. That's quite a lot of work, so let's reward ourselves with some fun facts about Bayes' Theorem.

Cool Terminology

Let's look at our basic equation yet again:

$$p(H \mid E) = \frac{p(H)\, p(E|H)}{[p(H)\, p(E \mid H)] + [p(\sim H)\, p(E|\sim H)]}$$

Mathematicians have labelled five terms in this equation:

- p(H|E) is the *posterior probability*; that is, the probability after the new piece of evidence (E) is accounted for.

- p(H) is the *prior probability* (or simply "*prior*" for short); that is, the probability before the new piece of evidence (E) is accounted for.

- p(E|H) and p(E|~H) are *likelihoods*. This is an unfortunate term because to most people "likelihood" is synonymous with "probability." But that's what they're called.

- [p(H) p(E|H)] + [p(~H) p(E|~H)] is the *normalizing factor* (or *normalizing constant*). Mathematically its role is to make sure that the posterior probability has a value between 0 and 1.

It is not necessary for you to learn these terms in order to use Bayes' Theorem. But if you run into a probability nerd, you might want to throw some of these terms into the conversation.

"A rose by any other name ...."

In this book we have used the word "probability" to refer to *subjective* probability, that is, degrees of personal belief. Many statisticians maintain that this is an error and that the word "probability" properly refers only to *objective* probability, that is, the number of times an event occurs (or should occur) in a large number of opportunities.

I suggest we stay out of this arcane dispute[12] and adopt a pragmatic approach. Given that Bayes' Theorem is mathematically valid, let's use it whenever we have a prior in which we have some confidence—although even if we don't, we will see in the next chapter that the subjective probabilities of two people with very different priors can sometimes converge rather quickly as new evidence comes in.

When You're Absolutely Certain

Look closely at Bayes' Theorem:

$$p(H \mid E) = \frac{p(H)\,p(E|H)}{[p(H)\,p(E \mid H)] + [p(\sim H)\,p(E|\sim H)]}$$

If p(H) is 1.00—and p(~H) is therefore 0.00—then p(H|E) is always going to remain 1.00, regardless of the values of p(E|H) and p(E|~H). Similarly, if p(H) is 0.00—and p(~H) is therefore 1.00— then p(H|E) is always going to remain 0.00. This makes sense. If your subjective probability of H (or ~H) is 1.00, you are essentially saying that no evidence whatsoever could *ever* cause you to change

---

[12] There are many arcane disputes about Bayes' Theorem.

that subjective probability. We are talking now about a religious-type certainty. And sure enough, Bayes' Theorem bears this out.

KISS

The formula for Bayes Theorem used in this book—which is the one relevant to most of the problems you will encounter in everyday life—is used when (i) you deal with two hypotheses, H_1 and H_2, and (ii) H_1 and H_2 are opposites (technically, H_1 is the "contradictory" of H_2). If H_1 is true, then H_2 *must* be false and *vice versa*. Further, either H_1 or H_2 is true; that is, there is no other possibility. Mathematicians refer to this situation as H_1 and H_2 being *mutually exclusive* and *jointly exhaustive*. In this situation, we can call the two hypotheses simply H and ~H, as we have been doing in this book.

But what if the situation is more complicated? For example, suppose you assign a prior subjective probability to one hypothesis (H_1); a different prior subjective probability to a second hypothesis (H_2); and a third prior subjective probability to all other hypotheses lumped together such that the sum of all three subjective probabilities equals 1.00? Or you have three specific hypotheses and then everything else? Or four, five, or more? Even worse, some hypotheses are mathematical, come in the form of a continuous variable, and can be represented as part of a probability density function.

In these and other situations the mathematics of Bayes' Theorem rapidly gets much more complicated than the relatively simple equation we have used in this book. That is why textbooks about Bayesian statistics usually look so forbidding to non-mathematicians.

In fact, Bayes' Theorem was discovered by Reverend Thomas Bayes (1701–1761), a somewhat obscure English minister and mathematician, and first published shortly after his death.[13] It took approximately two hundred years before computers gave us the computing power necessary to use these more complicated

---

[13] The famous French mathematician Pierre-Simon Laplace (1749 – 1827) subsequently discovered it independently but surrendered his claim of priority when shown evidence that Reverend Bayes derived it first.

formulations. But not to worry—we're not going there in this concise introduction.

Chapter Nine—"Washing Out The Priors"

It might seem that the results we reach by using Bayes' Theorem are very sensitive to our initial subjective probability— p(H). What happens if two people assign very different initial probabilities to a hypothesis? These are *subjective* probabilities after all. Nevertheless, those very different initial probabilities can sometimes quickly converge <u>if they agree on the likelihoods relating to new evidence</u>. This is called "washing out the priors."

For example, suppose two people assign initial probabilities to a given hypothesis of 90% and 20%, respectively. Watch what happens if they agree as to the likelihood of subsequent pieces of evidence—we'll follow the calculations for four iterations:

<u>Person #1</u>:

p(H) = 0.90

p(~H) = 0.10

<u>Person #2</u>:

$$p(H) = 0.20$$

$$p(\sim H) = 0.80$$

Evidence A:

$$p(EA|H) = 0.75$$

$$p(EA|\sim H) = 0.25$$

Person #1:

$$p(H \mid EA) = \frac{p(H)\,p(EA|H)}{[p(H)\,p(EA \mid H)] + [p(\sim H)\,p(EA|\sim H)]}$$

$$p(H \mid EA) = \frac{(0.90)(0.75)}{[(0.90)(0.75)] + [(0.10)(0.25)]}$$

$$p(H \mid EA) = \frac{0.675}{0.700}$$

$$p(H \mid EA) = 0.964$$

Person #2:

$$p(H \mid EA) = \frac{p(H)\,p(EA|H)}{[p(H)\,p(EA \mid H)] + [p(\sim H)\,p(EA|\sim H)]}$$

$$p(H \mid EA) = \frac{(0.20)(0.75)}{[(0.20)(0.75)] + [(0.80)(0.25)]}$$

$$p(H \mid EA) = \frac{0.150}{0.350}$$

$$p(H \mid EA) = 0.429$$

Evidence B:

p(EB|H) = 0.60

p(EB|~H) = 0.40

Person #1:

$$p(H \mid EB) = \frac{p(H)\, p(EB|H)}{[p(H)\, p(EB \mid H)] + [p(\sim H)\, p(EB|\sim H)]}$$

$$p(H \mid EB) = \frac{(0.964)(0.60)}{[(0.964)(0.60)] + [(0.036)(0.40)]}$$

$$p(H \mid EB) = \frac{0.5784}{0.5928}$$

$$p(H \mid EB) = 0.976$$

Person #2:

45

$$p(H \mid EB) = \frac{p(H)\,p(EB|H)}{[p(H)\,p(EB \mid H)] + [p(\sim H)\; x\; p(EB|\sim H)]}$$

$$p(H \mid EB) = \frac{(0.429)(0.60)}{[(0.429)(0.60)] + [(0.571)(0.40)]}$$

$$p(H \mid EB) = \frac{0.257}{0.485}$$

$$p(H \mid EB) = 0.529$$

Evidence C:

$$p(EC|H) = 0.70$$

$$p(EC|\sim H) = 0.25$$

Person #1:

$$p(H \mid EC) = \frac{p(H)\,p(EC|H)}{[p(H)\,p(EC \mid H)] + [p(\sim H)\,p(EC|\sim H)]}$$

$$p(H \mid EC) = \frac{(0.976)(0.70)}{[(0.976)(0.70)] + [(0.024)(0.25)]}$$

$$p(H \mid EC) = \frac{0.683}{0.689}$$

$$p(H \mid EC) = 0.991$$

Person #2:

$$p(H \mid EC) = \frac{p(H)\,p(EC|H)}{[p(H)\,p(EC \mid H)] + [p(\sim H)\,p(EC|\sim H)]}$$

$$p(H \mid EC) = \frac{(0.530)(0.70)}{[(0.530)(0.70)] + [(0.470)(0.25)]}$$

$$p(H \mid EC) = \frac{0.371}{0.489}$$

$$p(H \mid EC) = 0.759$$

Evidence D:

p(ED|H) = 0.90

p(ED|~H) = 0.05

Person #1:

$$p(H \mid ED) = \frac{p(H)\,p(ED|H)}{[p(H)\,p(ED \mid H)] + [p(\sim H)\,p(ED|\sim H)]}$$

$$p(H \mid ED) = \frac{(0.991)(0.90)}{[(0.991)(0.90)] + [(0.009)(0.05)]}$$

$$p(H \mid ED) = \frac{0.8919}{0.8924}$$

$$p(H \mid ED) = 0.999$$

Person #2:

$$p(H \mid ED) = \frac{p(H)\, p(ED|H)}{[p(H)\, p(ED \mid H)] + [p(\sim H)\, p(ED|\sim H)]}$$

$$p(H \mid ED) = \frac{(0.759)(0.90)}{[(0.759)(0.90)] + [(0.241)(0.05)]}$$

$$p(H \mid ED) = \frac{0.683}{0.695}$$

$$p(H \mid ED) = 0.983$$

Thus, two people started with subjective probabilities for a certain hypothesis of 90% and 20% respectively. Four pieces of evidence later they are at 99.9% and 98.3%, respectively. This suggests that the discussion between them should focus more on the likelihoods relating to the subsequent evidence than on the initial prior probabilities.

Conclusion

Although we usually don't focus on the fact, we constantly categorize the probabilities that our beliefs about things are true. We think some beliefs are almost certainly true; others are highly probable; and others perhaps just more likely than not. If pushed, we could assign a percentage chance—at least a rough one—that our various beliefs are true.

Hopefully, when we get additional information, we adjust the probabilities that we assign to our subjective beliefs. Far too often, however, we overweight the additional information and underweight (or even ignore) the older information that formed the basis of our original probabilities. Over time this can have serious consequences—unless, of course, we know how to use Bayes' Theorem, which you now do.

Printed in Great Britain
by Amazon

41939551R00029